This Other Kind Of Child

Are we harming our society by neglecting them?

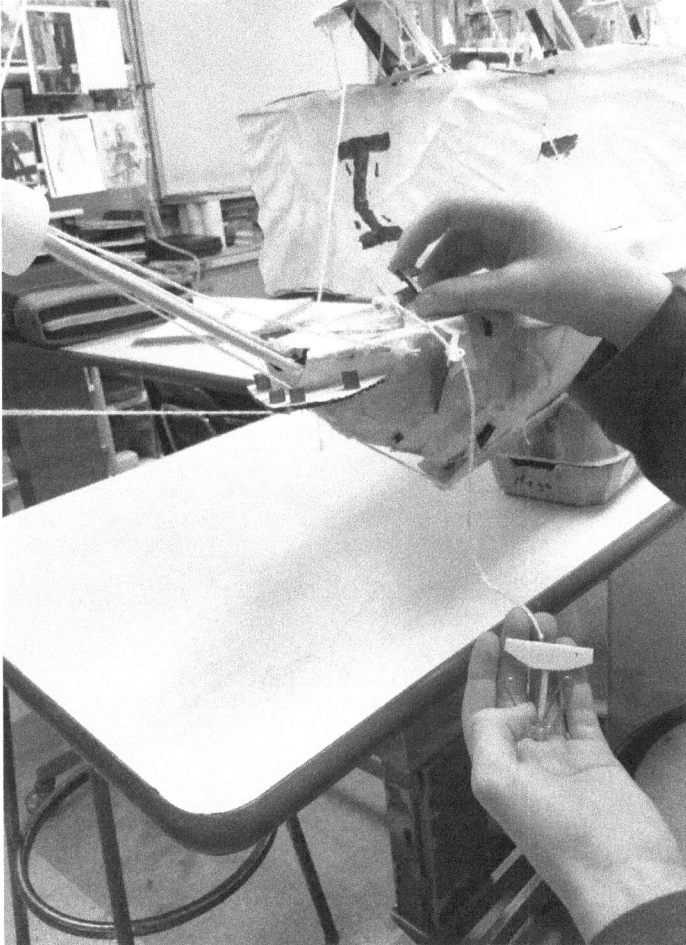

Meredith Olson, Ph.D.

http://www.DocOsBooks.com

Published by
SAN 299-2701
Glenhaven Publishing
4262 NE 125th Street
Seattle,Washington 98125

ISBN 978-0-9984627-3-8 (paperback)

Table of Contents

Introduction

A charming young man sat down next to me at the last faculty meeting. He looked tired. He had been working on admissions. Reading folders for the applicants to our school. He commented about how this was such a lot of work – trying to be fair to the overwhelming number of students wanting to enter our school. He had been told that in the past, our admissions process had paid particular attention to the Block Design sub-score of the WISC intelligence test, which all the applicants had taken. He wondered about the significance of the Block Design score. Did it really matter any longer?

Oh my goodness! That really grabbed my attention. Over 40 years ago when the fledgling school moved to this site, it was starting to develop an understanding of how Block Design scores help define the type of students we serve. We developed this understanding over a decade of observation and research. Somehow I have failed to pass on this understanding to the following generations of teachers. This writing is my attempt to correct that oversight. To help others understand what makes "this other kind of mind" different and precious. To help us all understand the things we do here to serve, nurture and protect these unusual students.

I have unapologetically devoted my professional life to trying to serve the population this young man was wondering about. How could I explain? It seems to me that these students with high spatial ability have been consistently and systematically ignored (and even oppressed) in modern education's focus on "mainstream" education. I taught in public schools for seventeen years and I watched them struggle. My urge has been to try to provide a safe haven at this school for these poorly understood students.

About 40 years ago I came to teach at Seattle Country Day School. I had already taught math and science in Bellevue, Washington and Seattle, Washington for seventeen years. I had raised four children and was beginning to notice temperaments, cognitive dispositions and personality variables. Some children were good scientists and puzzle solvers while others were good students who memorized science ideas. Having started out as a High School geometry teacher, I was particularly struck by the differences in geometric understanding.

Lucile Beckman, the founder of our school had noticed that the very brightest of our students had high scores on the Block Design subtest of the WISC intelligence test. That interested me. The test seemed to be tapping into some ability to notice geometric structures. I watched the test being administered many times and began to notice differences in children's approach to solving these geometric puzzles. I wondered what kind of spatial ability was being tapped into to achieve these amazingly high scores on the Block Design test.

I began testing other cognitive factors to see what children could do. I took kindergartners, one at a time, out into the hall and asked them to work a problem selected from the adult test battery I was exploring with our older students for my Doctoral research. I was amazed to find that many of our four-year-olds could solve the paper-folding challenge (fold, punch, unfold and match the pattern of holes) even though they had never folded paper snowflakes or paper dolls. How could they do that? Why could so many others not do it? Why were these the brightest children in school? How was this correlating with their reasoning? It didn't seem to have much to do with reading ability.

I kept coming back to Block Design scores again and again. Their behavior on paper folding correlated almost exactly with their Block Design scores. I wasn't getting new information. I was just supporting what I already knew. Block Design scores were the best identifiers after all. They seemed to be able to predict how well a child would do on other spatial tests I gave them. More importantly, the Block Design score seemed to correlate with the development of logic that began to show itself at the age of ten or so – at the end of grade four. We admitted most children in Kindergarten, before these logical abilities began to manifest in the developing human brain. How could we predict their behavior five years later? Did the Block Design really have that kind of predictive value?

More importantly, did the Block Design score predict who would become disenchanted with regular school? Could we predict which children would have needs less and less well served by mainstreamed curriculum? Was there a way to identify "the disadvantaged gifted"? If these were the Einsteins, the Feynmans, was there a way to protect them? To provide a safe haven for them? What kind of lessons would they thrive on? Be inspired by? Could our school become the bellweather "watering place" for this kind of valuable but misunderstood child?

I became interested in the physiology of the brain. How did brain structure relate to learning preferences. I conducted a number of studies videotaping children as they tried to solve spatial puzzles or to work problems on the Differential Aptitude Test. I noticed a difference in the hand they used to point to answers on verbal vs. spatial problems. My camera was able to pick up the way they tilted their heads suggesting a preference for visual field usage on one type of problem or another. Several of my studies were published in gifted education journals.

Cerebral lateralization in science
MB **Olson** - **Gifted** Child Quarterly, March 1,1979 – vol 23, issue 1, 1979

A study of the Correlation Between Verbal and Spatial Information Processing Modes and Piagetian Levels in Gifted Children. Meredith B. Olson Proceedings of the Seventh Interdisciplinary Conference (vol 11) Piagetian Theory and its Implications for the Helping Professions. UCLA,1978

The Impact Of Conceptions Of Giftedness On Curricular Design, Meredith Olson, Research and Issues in Gifted and Talented Education, ERIC Clearinghouse Feb 1981, No SP 017 369

What Do You Mean By Spatial?, Meredith B. Olson, Roeper Review, April 1984, p240-244

The time was ripe for investigating the contrasts more deeply. Being involved in a graduate program at the University of Washington, I had an interest in educational research. I was able to base my Doctoral Thesis on the amazing and unique population found at Seattle Country Day School.

Cognitive Styles in the Gifted Science Classroom by Meredith Beach Olson, Nov 27, 1984 University of Washington

After several years of focused observation of student behavior, as I gave assignments and various cognitive tasks, I had begun to discover relationships between students and lessons. A great many of my students had not been well served in prior schools. Why? Why did these bright students resist doing the standard assigned lessons? Why did several come to really dislike school? One would think that bright students should easily do well in school. Where was the mismatch? I could tell the students wanted to learn, but not the way they had been taught in their public school experience. It became clear to me that the really important thing for me to do was to try to match lesson design to the cognitive abilities and preferences of my students.

This "other kind of child" is not easy to teach. They fight with their lesson requirements. They get "fed up" with school. They may barely comply with school demands or they may simply drop out. What is the problem?

Over the years our school has grown. I now teach children of my former students. Alumnae report fond memories of the time they studied with me. They thrived on the unique instructional style I developed to address their temperaments and cognitive preferences. Even today, students in our population show dislike for math and science lessons taught in ways commonly found in mainstreamed classrooms. What is different about lessons that inspire them? Are the lessons easier? More difficult? Or is there some other lesson attribute that grabs their interest? How is that done?

Can this school remain a beacon of safety for this "other kind of child"? A haven for them to thrive in - as they grow through their unusual intellectual childhood? Can this school continue to develop lessons to provide a safe place for this unusual kind of child to develop their potential?

I have recently published a series of short works outlining the general flavor of lessons that work with these students under the title, *Raising Interesting People*. I have also set forth some thoughts about the relationship between scientific discourse in my classroom and the development of thoughtful citizens under the title, *Never Trust A Science Teacher*.

This volume contains appendices about how I discovered thinking patterns in the children I teach and the lens I use to design lessons that energize them. Included here are also some vignettes which may provide a window into the mind of this "other kind of child".

The issues discussed in the appendices are:

1. How can we recognize "this other kind of child"? What are they like? In college? In High School? In Elementary School?

2. How can we find them in a population of bright students? Is the Block Design sub-score of the WISC a useful identifier of "this other kind of child"? What does our research show?

3. How can we provide a safe haven for them? How can we better serve their needs so they want to fully engage in developing their potential? What teaching style works?

Children Differ

Children differ. They have different personalities and innate temperaments. Some are loudly social. Some are quietly thoughtful. Some play with dolls. Some play with trucks. Some read early. Some read late. The range of human behavior is large and important. Much of what a child will become is innate, but much is learned.

This book is about a special set of children. Hard statistical data (discussed in Appendix #2) reveals how our population of children differ one from another. We attempt explanations of why we see the behaviors we do. We include suggestions of how to engage them in school.

The most serious challenge educators face is the child's dislike of school. Although they love learning, these students often dislike assigned lessons. They are stubborn. They frequently fail to complete them. Teachers become exasperated. Thoughtless remarks and reckless insistence on completing lessons may lead to children dropping out of school. Parents may become fearful for the future of their child. And rightly so.

I have tried to follow the developmental patterns of "this other kind of child" for over fifty years of teaching. They don't display particular disabilities but are often willful and stubborn and focus on doing things their own way. These are the potential geniuses. They will become recognized as such if we can keep them involved in learning long enough for them to develop a knowledge base useful to their future and to society.

It may seem to the parents that their child is "the only one" of them. These children often search for years to really find a friend. A person who enjoys the world from the same vantage point. But what is that vantage point? That world view? How can we know what to look for in seeking to find a friend for them?

Parents notice when their children sit up, crawl, walk, and talk. They can't help but compare. They notice what children play with. Dolls and trucks. Dishes and sand. They are sensitive to gender stereotyping and usually try to make available all types of toys so children can discover their own self in a gender neutral way.

And yet, children differ. Some carry around toys. Some pile them up. Some youngsters are attracted to 2-dimensional jig-saw puzzles and others grab Rubik's Cube or Chinese Rings. Once one is aware of what to look for, one need only watch closely to notice persistent differences in the behavior of children.

Parents may have contradictory feelings about children who "take care of things" vs. children who pile things up, crash things, or take things apart. Which is better? Better for now? Better for their future? How can we know? Destructive behavior does not automatically predict future brilliance. Being willful does not necessarily predict our future engineers. But reaching for a 3-D puzzle once in a while, does. At least it does in the children I have observed.

These children often display musical or artistic talent. I have never encountered parents who relate music and art to a child's preference for 2-D vs. 3-D activity. This question doesn't occur to them. **Yet the difference between 2-D and 3-D awareness and interest is the central discovery of my research with kids**. It matters. It correlates with their willingness to patiently endure school lessons (or not). It dictates the kind of lessons that interest them.

Our study, described in Appendix #2, identified three types of students. *Type 1 students*: the conscientious, socially aware, non-puzzle oriented students showed strengths in 2-dimensional spatial recognition. *Type 2 students*: the puzzle acquiring non-original mathematicians showed strengths in mental rotation of

3-dimensional objects. *Type 3 students:* the creative puzzle obtainers revealed a superb ability in 3-dimensional spatial transformational thought. Type 3 is "this other kind of child" discussed here. **It is quite likely that an instructional mode which interests one individual may be found to be oppressive or "boring" to students with the opposite abilities**.

But the new setting has to be interesting. Not activity for activity's sake. It has to have intellectual challenge – appropriate to the age of the child. Static projects, such as those often assigned in schools, are often found boring by these analytical minds. Their attention may wander. Schools may blame the child – but the real problem may be the lesson.

When did these spatial preferences begin? Well, it can be observed pretty early if you know how to look for it. Piagetian tests of the *volume of juice in glasses* or the task of *expecting a train to appear at the end of a tunnel*, are solved by toddlers. By the age of 4 or 5 these children are building tall block structures, using repeated pattern elements, and trying out extremes such as cantilevers. The children I have watched take a liking to mathematics, chess, computers and often play the piano.

As they grow, many become interested in science fiction and physics. Science Fiction covers a wide spectrum from fantasy to engineering. In my experience, these unusual children prefer machines to magic. They love a good twist of a physical law. At the same time they are learning a lot about society and the nature of civilizations as they ravenously read about the mechanics of space travel. Who do they grow up to be? Some are talkative and some are shy. Their personalities cover a range. Some become successful mathematicians, physicists and computer engineers. A few have turned out to be "starving" artists. But all are interesting people with strongly worked out belief systems. These are people worth having in society.

The urgent problem is that these unique, strong-willed individuals may simply ignore whatever they find boring in school. They may make no attempt to master the very subject matter that could be so motivating to their logical development.

We, as educators, would be well advised to search for methods of exciting the curiosity of "this other kind of child" in the enduring knowledge base of our society. We must not assume that our young students are unable to engage in deep thought. By the age of ten or eleven, children can become passionately involved in studies of the *Earth As A Geodynamo* or *intricate micro-engineering of heart defects*. The educator's task is to go beyond the easy, superficial topics and find a way to engage students in profound and enduring understandings that drive our technological society forward.

The bottom line is, "this other kind of child" is a major part of what is distinctive about this school. It took me more than a decade of observation and research to develop an understanding of "this other kind of child". Serving this "other kind of child" is what we do here. We are not serving a privileged group, we are serving an oppressed group. These children who visualize 3-dimensions and use logical reasoning even with nonsense syllogisms, are rare. Their emotional sense of self-worth is assaulted on a daily basis in mainstreamed curriculums. Typical school classrooms emotionally oppress this child. Our school is a singular hope for them. We have an opportunity. The Block Design score matters. It is the best predictor of this unique kind of child who we seek to serve. This underserved kind of child needs us. The need is urgent. These children are worth saving. They are the future of our technological society.

Appendix #1: Anecdotes

What is this "other kind of child" like?
How do they think differently from most students?
We need to characterize them so we can serve them.
We need to better understand who we are looking for.
Anecdotal snapshots provide a window into their minds.

"By their words ye shall know them"
European Journal of Social Psychology, April 16, 2009 Douglas & Sutton

Here are some examples.
    A. From college age people.
    B. From high school age people.
    C. From elementary age people.

Thoughts They Had In College

Essays written by "this other kind of child" when they were college age leave a paper trail of their thinking. These excerpts are included to give a window into thought patterns of these young adults.

An Essay

A delightful parody written by someone with this kind of mind.

Rich Isaacman is currently vice president of science and engineering of ADNET Systems, a firm that supplies technical expertise to NASA. He received his Ph.D. in astrophysics in 1980. Four years earlier, this article came out in *Analog Science Fiction* magazine. What kind of mind thinks like this?

Einstein wrote about *Special and General Relativity*. Isaacman *(Analog Science Fiction/Science Fact, January 1976 p164-167)* wrote about "*Special and General Creativity*". His essay begins with:

The astute student of physics, having up to this point been introduced to such idealistic sounding laws as "Conservation of Energy" and "Conservation of Matter" should, in preparation for future work in a scientific field, become acquainted with one of the harsher, though more immediately practical, laws of the natural and academic worlds. The law is "Conservation of Status," or, as it is also known, "Publish or Perish."

The essay begins innocently enough to intrigue literature teachers. It points out that Einstein achieved immortality by writing exhaustively about only two highly theoretical and largely impractical theories. So far, so good. Isaacman then goes on to derive mathematical equations for "Intellectual inertia" (which he calls sluggishness) measured in units of "Lethar Gravities," (or Lether G's) named after Edward Lethar, who it is noted, died in obscurity because of a failure to get any work published.

The essay derives an equation for "creative acceleration" (words per second) which turns out to be inversely proportional to Sluggishness. Not satisfied with that, the essay continues to examine the limit (mathematical limit, that is) at which the velocity of words remains constant. A further excerpt from the paper provides a clearer idea of its style and content.

"It is clear that in order to produce a creative acceleration, there must be some force acting upon the Sluggishness of the writer. The resultant acceleration will be proportional to this force, though inversely proportional to the Sluggishness. Thus

$$A=F/S \text{ or } F=SA$$

Acceleration is measured in "words per second squared.".
But just what is this force? Clearly, it must be external in origin, since in general, all internal forces cancel by symmetry. The standard representation of a force, as when dealing with fluids, is the product of pressure and area. In this instance, the pressure is the external creative pressure on the writer, measured in the case of a student writer in "grade penalty per day late," or for the case of a freelance writer, in "cents per word paid." Area is measured in "pages" or square pages. (NOTE: The international standard "square page" is defined as being 8 ½ x 11, which actually isn't square at all.

Example: If a student laboring under one Standard Earth Sluggishness (9.9 Lethar g's) has a ten page paper due with the teacher imposing a penalty of half a grade per day late, what is his creative acceleration?
Solution: External Force = pressure x area

$$F = PA$$

Creative acceleration a = F/S

$$=PA/S$$
$$= (.5) (10)/9.8 \text{ words/sec squared}$$

Hence the student will write with an acceleration of approximately 0.5 words per second squared.

Let us examine this result. According to the rules of kinematics, velocity v=v + at, that is, with an initial velocity of zero, the velocity at time "t" of a body undergoing an acceleration "a" will be equal to the product of that acceleration and the time the body has been undergoing it. Therefore, at an acceleration of 0.51 w/sec squared after a period of only 100 seconds, or one and two-thirds minutes, the student will be writing at

*a rate of 51 words per second. This is, charitably speaking, highly
unlikely for a human being. What has gone wrong?*

*CREATIVITY RELATIVITY: What has "gone wrong" in the previous
example is that we assumed all quantities to be static and unchanging.
In fact, this is not the case. Because of the necessity of having to do
more work as one writes faster and faster, the Sluggishness of the writer
increases with velocity ...*

*From the above, it can be seen that the value of "c" is thus the ultimate
velocity of the writer, since creative acceleration must cease when it is
attained. It is therefore referred to as the "Velocity of Write."*

The paper continues for several more pages discussing momentum
and energy, friction, distraction in terms of six more derived
equations. The paper ends with a section entitled *CONCLUSION.*

*We have concluded our introduction to the fundamental principles
underlying creative writing. It is hoped that, equipped with this
information, the student of Physics or English will be able to transcend
mere knowledge of his field, and proceed to a higher state of
understanding in which he can convince others that he knows what he is
talking about regardless of whether that is actually the case. By
mastering Creativity, the ratio of constructive action to abstruse thought
can be increased to its theoretical limit, toward the ultimate goal of
writing endlessly on a new idea which itself took only a few seconds to
think out. Therein lies the secret of success in the academic community.*

What literature teacher would be willing to score a paper like that?
Clearly this essay contains structural analysis of both literature and
mathematics. This writer is concerned with previously
undiscovered relationships between the style of mathematical
proof and the creative interplay found in literary works. Using the
"conventional language" of mathematical discussions, the author
has interposed the subject matter of literary creativity. Intentional
alterations in foreground and background are obvious. We see a
mind at work which is equally comfortable with literary style,

humor (derived from double meanings and homonyms from the world of physics) and mathematics. The author is so comfortable with multiple modes of thinking that he is able to play with them and intertwine them creatively. Unfortunately most literature teachers are not able to read the mathematics, let alone recognize the way equations parallel (and lampoon) famous equations in physics. A student of this type quickly learns that the world of literary composition is restricted to a very narrow channel of thought (from his view). This type of student will usually conform to the demands of a literature teacher for most of his high school career in order to make it through the "system". Although the compositional skills demonstrated in this essay are exactly those most highly prized by English teachers, this type of humorous critique of literary style is normally revealed only to kindred spirits outside of the school setting.

A Rhodes Scholar Application

Who are these people? These "other kind of people"? What are their values. What kind of people do they become? In his personal essay for the Rhodes Scholarship Finals, this "other kind of child" made the following statements:

During fifth grade I acquired my first intellectual activity, one that has stayed with me to the present: the reading of science fiction. At that time I discovered a series of books entitled Tom Swift. I was fascinated by this series because each book was built around a clever idea, usually embodied in some sort of machine. In the course of resolving the plot, the machine was used in some highly creative and unexpected way. These and other books stimulated in my mind innumerable questions of "What if...?" I would dream my way from one creative solution to another. To this day I retain my interest and delight in "What if ...?" types of questions, and I fervently hope I always will, for I feel it is one of my best and most important personal characteristics.

However, as I have matured I have discovered that there are many, many things in this world that are interesting in addition to machines. By the end of high school, I had developed strong interest in math (all forms but especially geometry), physics, computers, linguistics, and philosophy, and had spent a good deal of time discussing aspects of these topics with my friends.

For a variety of reasons I decided to study science first, specifically physics. To this end I came to Caltech. Here I have energetically pursued my interest in physics, establishing, I feel, a strong enough foundation that I can effectively work in any part of physics without more formal class instruction.

Caltech's courses in market theory and game theory are excellent. They prompted me to toy with "ideal" economic systems and with ways to improve existing systems. In addition, during term breaks and summer vacations I spent a great deal of time considering topics in education, such as, how do people learn, how should one teach, how does one develop a personal ethic, how does one develop logic skills and creativity, and especially how can one help others develop these qualities and skills.

At this point one might well wonder what the point of all the studying is, what is it leading to? A simple and straight forward reason is that I enjoy learning. I enjoy diverse topics, intertwining them, synthesizing new ideas from them and making "what if ?" questions out of them.

Pragmatically, such a broad range of interests is necessarily in order to prepare one to deal with the world as it is becoming. One needs to have spent a good deal of time contemplating moral and ethical questions so that one can wisely decide how to use the technology we have created. Without both one will not be able to deal effectively with the problems that our technology will create.

Finally, I believe that an ever questioning, ever learning approach to life will help me develop the compassion and wisdom that I have seen and admired in such a man as Jacob Bronowski. It will help me develop the compassion to understand to some extent those people around me, how they feel, their needs and wants. It will help me develop the wisdom to see the vast interconnectedness of the world so that I can understand the consequences of my actions and thus be able to help others without damaging them, but on a personal level and at the level of guiding (in a small part) of future human development.

A graduate school application reveals the kind of person this "other kind of child" is becoming.

"How do ALL things work and why do they work in whatever way they do?

This question is the cornerstone upon which I have, am, and will try to build my life. Unfortunately, there does not seem to be enough time in my life to deal explicitly with the how's and why's of ALL things in the way I would like to. Hence, I – sigh – have to be happy with merely studying those things that explain the how's and why's of as much as possible, as efficiently as possible. Now, making the most general, most fundamental statements about things seems to be the province of Physics. Hence I have, am, and will study Physics.

My current interest is the application of methods of differential geometry to basic problems in the areas of general relativity theory and/or high energy-density theory. In particular I wish to use this formalization to clarify the basic problems in these fields and as a guide to effective methods of solution. (eg. Local and globe transformational symmetries of the theories, breaking of those symmetries and geometrical interpretation of what is happening, Lie-Backlund transformation techniques, Phase space analysis limiting behavior at large times)

I am also interested in using other techniques (eigen-function expansions, asymptotic expansions w.k.b., weiner hopf, etc) in the above and following fields; thermodynamics, statistical mechanics, phase transitions, low temp, solid state (particularly quantum electronics and quantum optics), plasma physics (fusion and the problem of muon catalysis), and anything else.

The particulars of this are completely open. My past experiences have been primarily with scattering type experiments, either nuclear or high energy. I helped Dr… perform light nuclei fusion experiments. I worked with Dr… in the Tasso group at Petra studying an e e colliding beam storage ring at the Deutches Electron (DESY) in Hamburg Germany."

This "other kind of child" often (usually) develops into a kind and thoughtful adult. Their world-view is revealed in the applications they write to graduate schools. It is interesting to note that their consideration of morality and humanity grows out of their interest in physics. Physics first, humanities later. These students are resolute in their interests and will not conform to our current curricular expectations which require analysis of morality and humanity **instead of** an analysis of physical laws. We need to find educational opportunities that recognize their needs. Many of the students with whom I have worked show an interest in physical science between the ages of ten and sixteen. By the age of seventeen, many of them have expressed an interest in philosophy. We want this kind of citizen. What can we do to nurture them?

Thoughts They Had In High School

High School age is tumultuous. Students are carving out their own individuality. They are becoming who they are. They don't always let us know. They are developing a private world of their own even though, on the surface, they might seem cooperative and compliant.

These essays are included to try to capture the struggles and strivings of High School age children.

Two Young Men

Two young men I once knew developed the habit of going off together on Saturdays. One was a sophomore in high school at this time; the other was a junior. They would leave about 9 a.m. and return to their homes in the evening. Being good students and responsible individuals, no one paid much attention to where they went or what they were doing. Their activities were eventually revealed in the course of casual conversation. It seems that they had developed a game which they called "Random Library". They would to the the library of the local university to play. As they explained it, their major problem was developing a method for generating truly "random" numbers. This often took them most of the morning. After they were satisfied with their set of "random" numbers, they looked up the books registered under these numbers in the large card catalogue. After collecting all of the books identified in this way, they spent the rest of the day finding relationships between and among the selections concerning philosophical, mathematical, and literary ideas. They examined the interplays of literary background and foreground, the ways in which traditional grammatical structures of conventional language had been intentionally violated by the authors, and the methods by which such creative violations could be produced by intersecting one randomly chosen work with another. They were examining not only what the passages meant but how the language operates. This was direct inquiry into structural analysis of literary composition. Mathematical operations were as acceptable as linguistic operations. Rhyme schemes and linguistic flow, time and continuity, climax and resolution, sequence, series and the mathematical concept of limits all were woven into the entertainment of the day.

Very few people ever found out about this activity. It was done for the humor and intellectual satisfaction. None of it was written down, of course. None of their school teachers ever learned of this activity. Both young men became National Merit Scholars and entered PhD. Programs in physics at different universities.

The First Young Man

The first young man progressed through his high school career in
literature with "straight A" grades, but without sharing his personal
interests. He taught himself to type and to spell in a moderate way.
He frequently commented on his great enjoyment of his year of
philosophy. Looking through his compositions in his pile of
school papers produced the following list:

A Look at Pride and Prejudice *in terms of its title.*

*Alice in Wonderland: a commentary on the way Victorian society
stifles people's minds.*

*"Only he who does what is right acts freely." Jaspers
An argument against Plato's definition of freedom to do wrong.*

*Happiness (your own and others) is not the end at which the good
man aims; rather it is the reward which he receives.*

On Plato's Theory of Education: questioning the absolute good.

*On Time: (Augustine sees the world as a phenomena of the
present. I see the world as a four-dimensional continuum through
which we move)*

*Descartes and absolute truth (Problems with the definition of
existence)*

"What shall I conclude from all this evidence?" (on Descartes)]

The nature of physical reality

*Communication is what makes man human. (on Jaspers)
(Communication is essentially a sharing of experience and if
people don't have common experience they can't communicate.)*

Many of his papers were over twenty pages in length. All were well documented with quotations and footnotes. This student obviously had strong language art skills, and yet he frequently was at loggerheads with his language arts teacher. He and his friends could be overheard making disparaging comments about literature courses as they worked at computer terminals of the school. The following quotation from one of his essays points out one of the problems:

"I have read the first three and one-half chapters of "A clockwork Orange" and don't like it. I see no point in reading about sex, violence, and other things of low morality because I believe I am what I get from my environment.

I can see two grounds on which you could base an argument for having me read this book. One is that this book portrays some of the conditions that have happened in England, that are happening now, and that may happen in the future, and I should not blind myself to them. I agree, but I would prefer to find out about these from facts – not fiction based on facts. The other point is that many people read and like this type of stuff and that I should read it so that I can find a common ground with them. After all, I am going to live with them for the rest of my life. This is true but being moral means more to me than being social."

I feel that an English department should try to raise the level of morality, not lower it. I feel that we should read good, thoughtful literature – not psychological trash. (This complaint is not against you. I had the same complaint last year.) Instead of assigning a little bit of all types of literature, school has given me 75%trash and 25% Shakespeare.

To this his teacher replied:
"Did you ever consider how much violence and immorality is depicted in Shakespeare?"

Most English programs in secondary school study historical or thematic categories in literature. Literature teachers seem to emphasize how the literature is surrounded by the world of social actions and the world of individual thought and ideas. (Frye, 1973) Most questions asked by teachers emphasize the sociological and psychological aspects of the story. Before reading a selection, students are frequently asked to focus on vocabulary words, to identify the type of person who they consider a hero or villain, and to consider a focused type of friendship or social setting. Questions and tasks to be done during the reading frequently require active response to help the student focus attention on the crucial aspects of the selection – again most often from a psychological perspective. Questions such as, "Why did people react that way?", "What do you think they will do?", and "What is the author's perception of (x)"?, all typically force the student into psychological analysis. Even the exercises required of the student after reading the selection which are viewed as moving toward syntheses, analysis, and evaluation, typically focus on social science aspects rather than mathematical or physics related ideas. Students are asked to describe the cultures from which characters in the story came, to think of other interactions which might have brought the story to closure, or to analyze plot, characterization, theme or symbolism.

The very common focus of literature classes on the negative social interactions of society runs counter to the interests of many young men and women who wish to focus their free inquiry time on physics, technological innovation, mathematics and philosophy. I have noticed many students of this type who are willing to read classics such as Shakespeare and Chaucer which contain violence, but which have internal structure which is richly interwoven. They do not dislike literature, but almost without exception, they find great displeasure with the way literature is presented in the typical English classes they encounter. Literature instruction does not meet the needs of this "other kind of child."

The Other Young Man

Being a brilliant and willful individual, this young man frequently found himself outside the expectations of society. Because he knew how to work mathematics problems, he frequently didn't bother to finish them. The remainder of the proof was "too obvious" to be of enough interest to complete. Not a way to win favor with a math teacher. Never-the-less, this student was willing to complete written compositions because, he said, they offered more complexity to deal with than simple equations.

As a junior in high school, he grew tired of the course offerings available, and abruptly left school at the end of his junior year. Since he had completed all of the graduation requirements anyway, (including advanced calculus, physics, chemistry, biology, history, three years of foreign language, and language arts), he was quietly given a High School Diploma.

The next year he enrolled in the local college while he decided "where he should study" – that is, to which universities he should apply for regular college education. As it turned out, he was accepted by many of the Ivy League colleges, was awarded a large scholarship for academic excellence, and graduated after three years of "challenging" courses by simply passing competency exams. He majored in theoretical physics.

His first term at the local college was found boring. The next term found him demanding the right to take graduate level courses in Russian history, history, and philosophy. He also continued with senior level math and physics. Comments in a letter to his friend reveal a great deal about the thinking process of this young man.

"My classes this term are the best I've had at the U. In fact, the worst class I'm taking is the honors class in war and it was supposed to be the best. I think that points out the superiority of graduate classes.

Still, back at the war class, we have endless discussions (primarily because none of the textbooks we were supposed to read came in on time. One arrived yesterday.) The discussions are frustrating because of the massive cloudbank that occupies the minds of most of the class members and the professor. For some reason, God knows what, these people are hung up on the "a priori" assumption thet war is bad while at the same time they are trying to argue about how wars begin and how people behave in them. Their assumption is so overriding that it seems they can have no conception of the possibility that some people in war act on the assumption that war is not bad. Consequently, we are endlessly tangling ourselves in what appears to be contradictory behavior.

In support of our class, I must say, however, that there are very few sheep. I must say I was quite surprised at the complete blanket rejection that is given to many ideas brought up in class. (Of course, I rank high among those continually rejected.) It has reached a point where when I begin to make a point, it seems like I am talking to a wall – a wall that is continually being buttressed from the other side at that. When I realized this, my interest in the class greatly increased.

I can make the most solid of walls listen, so now it is a challenge for me to break down this psychological wall constructed against outside influence. My previous attitude was indifference. I think I will take a more outgoing posture in the future. This wall is significant in that it shows there are still people who stick to their own beliefs and who have the strength to hold them. Alas, as with most things of beauty, one has to test oneself against it by trying to tear it down.

There is a whole philosophy about "the wall" that I might tell you about some time. I'm being presumptuous, of course, but I think I could tell you some interesting new things about it.

About this time, (just after his seventeenth birthday) he decided he needed a job. Can you imaging what type of job he landed? Did you ever wonder who writes those Master's and Doctoral thesis on the "black market"? His letter commented:

"The big news of mine is that I have gotten a job of sorts. My job is well suited towards my demonstrated abilities. My demonstrated abilities being, of course, getting good grades. To end the suspense, I will tell you what I am doing. Quite bluntly, I write research papers for a company at $3 a page. I made nearly $50 in my first week (last week, by the way). These papers are on almost every conceivable subject. It appears that I am participating in the academic black market by writing these papers. (Exciting, isn't it ...) Ethically, I am unbothered though others in my place might be. To me, I'm doing what I've always wanted to do: be paid to study."

Thoughts They Had In Elementary School

Elementary school children still do not know their uniqueness. They are told they are "good" or "bad." It is all the judgment of the adult. This other kind of child does resist our rules and regulations on occasion, but mostly they comply. Yet, who are they? Who are they becoming? When do they resist us? What can we do to "strengthen their strengths" as we send them to school to "strengthen their weaknesses?"

Younger Children

What kinds of problems are encountered by this type of child in the elementary years? One child had a frequently encountered problem. The local library sponsored a "summer reading" program. Students were to read twelve books to be awarded a gold seal for reading excellence. This child (a boy) enjoyed the opportunity for summer reading very much. He would go to the library about once a week to check out fifteen to twenty books. He discussed his reading with the Librarian and obtained her signature for his completed selections. By the end of the summer he had registered over fifty selections with the Librarian. He was reading above his grade level and enjoying it immensely. He was in the third grade and was attracted not only to the Danny Dunn series, but to Tom Swift, Doc Savage and Lucky Starr. At the end of the summer he was informed that he did not qualify for the reading award. He had read in only one area of literature and the certificate required that he read in four areas (such as fiction, biography, poetry, fantasy and mystery). The child refused to read in the other areas. He explained to the Librarian that if this was indeed a program for pleasure reading then it should allow people to read as they chose. The child's name was not posted on the wall of the Library as one of the able readers of the community who had participated in the summer reading program, even though he had been reading nearly a book a day. Although, in the end, the boy was quietly given his gold star to take home, he was made to feel that he had done something "bad". It is no wonder that this student refused to fill out book cards indicating recreational reading for his classroom teacher the following year.

The Chevron Traveling Creativity Lecture

A few years ago, the Chevron traveling exhibit on Creativity
provided a speaker for an assembly in our school. The
presentation was made in the lunchroom. Children in the audience
were ages 7 to 14. As a teacher, I dutifully positioned myself at a
table filled with wiggly little boys. The speaker presented a
number of projects in which a rather large audience (about 80
students) participated. I remember one project in which IBM cards
were distributed to all the students. They were asked to create
something they might find useful to have on the moon. They could
fold, but not tear the card. For another task, they could tear the
card to produce a useful object. After a while the speaker asked
the students to put the cards down so she could go on to the next
activity. Some of the boys near me continued to mutilate their
cards in a purposeful manner. The speaker began explaining how
several great artists, scientists, and thinkers of past eras had refused
to conform to the expectations of their day. They had gone beyond
the guidelines given them by school and the scientific community.
She had to stop her speech several times to ask wiggly students to
put down their cards and listen to her as the time period for
working with cards was over! Spatial creativity?

My most vivid memory of the presentation was when the speaker
told two stories. The first story was told in good "story-telling"
form. Most of the children listened attentively. The little boys at
my table were not very interested. They quietly went "brumm,
brumm" with their pencils as they moved them like cars around the
table. I noticed that the students near me consisted primarily of
those who had been identified in my research as having spatial-
transformational ability. My next awareness was that the students
near me had begun to pay attention to the speaker. Several of them
were standing up, craning their necks to see what the speaker was
holding. At about this time I noticed many students in the rest of
the audience quietly drop their chin on their hands or even put their
head down. A complete reversal of attention was being exhibited

by students in the room. Those who had been attentive were slumping in their chairs. Those who had been "off in an imaginary world" were actively attentive. What had happened?

The speaker was holding up geometric shapes which were about 18" square. There were three shapes, a square, a triangle, and a circle. She was positioning the shapes in a sequence of arrays indicating the stages in the development of the story. She challenged the students to recognize the story she was portraying. The previously disinterested "spatially gifted" students were obviously having a grand time trying to decode the events. Finally someone identified the story: *Millions of Cats* by Wanda Gag.

We have seen that the problems of "this other kind of child" may stem not so much from their original lack of interest in language arts as from the difference between these students and their teachers. These students delight in intertwining spatial, mathematical and linguistic ideas. Many teachers are not prepared to deal with these.

Appendix #2: Research

What behaviors give a clue?
Look around a classroom. Can you notice them?
You probably won't see their abilities in regular lessons.
Assign a different task. Watch as the students separate.
What cognitive abilities do they possess?

Forty years ago we discovered that the Block Design subscore of the WISC intelligence test correlated with student behaviors and preferences about 75% of the time.

Additional research projects explored which cognitive factors the Block Design was eliciting. We found that 2-dimensional spatial ability was quite different from 3-dimensional spatial ability.

This provided a lens into how we might design lessons that would engage these populations of children.

The Subjects And The Setting

Seattle Country Day School had, due to involvement in research on cognitive attributes of the gifted since 1975, been particularly receptive to the admission of students with high Block Design scores on the WISC I.Q. test. This WISC subscore had appeared to the staff of the school to have some predictive value in identifying creative mathematics students. (Beckman, 1981) The sample population achieved scores in the upper half of the possible score range of the Block Design subtest of the WISC I.Q. test. Approximately half of the subjects scored between 16 and 19 on the Block Design subtest, while the other half scored from 12 to 15. The score range of the Block Design subtest is 1 – 19 with average ability being represented by a score of 10. Thus the "high ability" range from 12 to 19 was represented throughout its continuum by this sample population.

My study, which examined the "factor-specific" component parts of spatial ability, began in 1979. This study examined an intact population of gifted students enrolled in Seattle Country Day School in Seattle, Washington. Gifted middle school students (primarily grades four through eight) constituted the population under consideration. The WISC I.Q. test had been administered to all of the students in this study as part of the entrance examination for the school. All of the 62 middle school students were involved in this study. All scored above 125 on the WISC, with a mean score of 137.

The students involved in this study were plainly not a representative or random sample of all gifted students. However, since this study was focused on "spatial" modes of thinking, these students seemed to be particularly suitable.

Block Design Predictive Value

Seattle Country Day School has a forty year history of trying to serve as a safe haven for "this other kind of child." How do we find them? How do we serve them?

All applicants to the school take the WISC I.Q. test. We noticed that the Block Design subtest seemed to do a good job of predicting student behavior on the spatial tasks we assign.

Our original director put forth a lot of effort to correlate Block Design with classroom behaviors on a variety of tasks. Using an observer coding system she classified 7 to 9 year old students according to their perceived spatial functioning in three mathematics classes at the school. She derived the categories of high, average, and low display of spatial functioning and contrasted that with Block Design scores. She found that Block Design scores predicted spatial classroom behavior about 75% of the time. She published her findings in several educational journals. For several years we explored grouping students according to the spatial functioning behavior one period per day in order to assess the matching of instructional style to their enthusiasm. In this way we began to create lessons suited to our unusual population. To this day, alumnae report fond memories of the projects they participated in.

Data tables derived by Lucile Beckman show about 75% predictive value:

student	spatial high	functio average	ning low	BD 19-17	BD 17-14	BD 14-12	CORRE YES	LATION NO
1			X			12	X	
2		X				12		X
3		X			14			X
4	X			17			X	
5	X			19			X	
6	X			17			X	
7		X			16		X	
8			X			8	X	
9			X			13	X	
10		X			16		X	
11			X			14	X	
12		X				13		X
13	X			19			X	
14		X				12		X
15			X			13	X	

73%   prediction

student	spatial high	functio average	ning low	BD 19-17	BD 17-14	BD 14-12	CORRE YES	LATION NO
1	X			18			X	
2	X			17			X	
3			X			12	X	
4			X			12	X	
5		X			16		X	
6	X			17			X	
7		X				13		X
8		X		18				X
9	X					14		X
10			X			12	X	
11	X			19			X	
12			X			14	X	
13			X			13	X	
14		X			16		X	
15	X			19			X	
16			X	17				X

75 % prediction

	spatial	functio n	nin g	B D	B D	B D	CORR E	LATIO N
studen t	high	averag e	low	19 - 17	17 - 14	14 - 12	YES	NO
1			X			13	X	
2			X	17				X
3	X			17			X	
4		X			16		X	
5		X		17				X
6			X			14	X	
7	X			18			X	
8	X			19			X	
9			X		15			X
10	X			19			X	
11	X			17			X	
12		X			16		X	
13		X			16		X	
14	X			19			X	
15	X			19			X	
16	X			17			X	

76% prediction

About a quarter of our current population has Block Design scores of 18 or 19. Most of the rest are 15 and above. We recognize that some of these students may be difficult to teach. Our hope is that, through paying careful attention to the behaviors of students, we can increasingly learn how to engage and nurture "this other kind of child".

My Dissertation Use Of Tests

The design of this study called for the identification of students with similar profiles on factor-specific components of spatial ability. A search of the literature revealed batteries of "spatial" tests in civil service exams., military service exams, and differential aptitude exams. In my previous exploratory research on other populations of gifted students, the Iowa Cognitive Ability Test Battery and the Structure of the Intellect (SOI) had been used. Findings of these projects led to the theoretical expectation that certain types of tests might be more applicable to this research project than others would be. Some tests asked for quick recognition of an object which matched a given object. Other tests required that the subject rotate images mentally to determine the label on the hidden side of the object. Still others asked for assembling objects from the printed component parts. All tests consisted of typewritten words or black and white line drawings.

Some test authors claimed that their tests were discrete or "factor pure" when examined using various statistical analyses. While validity assertions of test writers were used as a general guide to test selection, it was felt that these gifted students might use thought strategies which would alter the definition of "factors." Several tests designed to examine a single "factor" seemed to elicit quite different response patterns in various gifted individuals. The purpose of this study was not to isolate separate "factors" unique to gifted minds, but simply to determine whether students who had exhibited similar classroom behavior would achieve similar profiles on an assemblage of logical and spatial tests.

In order to establish criteria for grouping students (using test profiles), which could be replicated by other researchers, no "teacher made" tests were considered for use. Components of I.Q. tests which had to be administered by a psychologist were also not considered for this study. Only tests which were designed to be administered in groups, which had been used in other applications, and which might be available to other researchers, were used. All of the cognitive ability tests considered for this study had been normed on other populations and had scoring procedures well defined.

The tests had been designed with consistency in mind. The Educational Testing Service reported form-associated reliability coefficients of 0.7, 0.8 or 0.9 between the two half-tests of each test derived from several sample populations.

Simplified aptitude factors used as descriptors by the test designers are shown below.

Test	Descriptor
Hidden Figures	similar to embedded figure
Concealed Words	out of focus pictures
Cube Comparison	spatial orientation – rotation in 3-D
Card Rotation	spatial orientation – rotation in 2-D
Hidden Patterns	embedded "man-like" figure
Nonsense Syllogisms	logical reasoning – nonsensical
Form Board	spatial restructuring component parts
Hidden Words	four letter words random letter lines
Paper Folding	spatial restructuring fold & punch
Deciphering Language	transitive reasoning symbolic lang.
Inference	if-then conclusions
Copying	embedded – copy dot pattern
Letter Sets	find rule to relate 4 letters
Surface Development	spatial restructuring fold-up
Gestalt Completion	out of focus pictures
Diagram Relations	Venn diagrams, transitive reasoning
Calendar	complex set of sequential directions
Shape Memory	visual memory for 4 minutes
Locations	find rule for dot-to-dot array
Identical Pictures	perceptual speed match items
Following Directions	complex set of sequential directions

The order of test administration was determined randomly by selection from a shuffled pile of tests. Each test had two halves which were administered on separate days. Recognizing the limited attention span of children, ten to fifteen minutes at the start of each science class period for a trimester were devoted to the testing. Test directions were read to the whole class at the start of each test session. Sample test items were presented on the overhead projector and correct answers were pointed out. This was done to help students understand the test directions which had been written for adults. Students were given a countdown until the starting moment. Papers were then turned over and the time allotment which had been established for adults was followed. The tests were hand scored using scoring procedures accompanying the tests. Scores were not revealed until the end of the term in an effort to reduce practice affects in score achievement on the second test half.

Scored were analyzed for each student to develop an individual profile of strengths and weaknesses and scores were normed for the entire population. The most interesting findings came from examining groups of scores from students who exhibited similar behaviors in the classroom. Students were partitioned into three groups.

Partitioning

I had noticed that gifted populations in Idaho and South Carolina both contained some students with interest in three-dimensional puzzles. For the purpose of this study, we decided to examine scores of those who worked puzzles and contrast them with those who did not.

Students had been observed for the first 50 days of the school year as they entered the math-science classroom. Each day three puzzles had been placed around the room on top of book cases, on shelves, and on various ledges. The puzzles used were The Brain, Looney Loop, Rubik's Cube, Chinese Rings, Frypan Maze, Whipit, Pyraminx, The Snake, and the Missing Link. Although middle school students typically entered the classroom in a disorderly manner, certain students always noticed a puzzle. Some idly handled it and some actually tried to work it with focused attention.

We called students Type A if they were never observed by a rater to try to work a puzzle. We called a student Type B if they were observed to grab a new puzzle as soon as they could and focus attention on trying to work it. Over 50 days, certain trends remained constant. In addition to attention focus, our middle school students were classified according to their behavior on inquiry math problems. When invited to explore Magic Squares, Pascale's Triangle, or Color Cubes, students who made original discoveries were classified as Type R while those who failed to make their own discoveries were labeled Type S.

It was found that Type R students all were Type B students. Type S came from both Type A and Type B students. This allowed us to form three groups of students. We partitioned the test score answer sheets into these three groups of students. We called A-S Type 1 students. B-S students formed group 2. B-R students were classed as group 3.

Type 1 was non-puzzle focused and non original on math projects.
Type 2 was puzzle focused but not original on math projects.
Type 3 was puzzle focused and was original on math projects.

We calculated mean scores for the three groups and for the middle school as a whole. We then calculation standard deviation of these groups for all tests. We found that the scores of group 3 was above the mean of other groups on most of the tests. Converting the scores of the 3 subgroups and the group as a whole to adult z-scores (which were available in the test direction packet) showed remarkable results.

Had we not partitioned scores, we could have concluded that our gifted population was best characterized by rapid sequential thought. By partitioning the scores, we found something quite different.

Type 1 was much higher on recognition and direction following skills. We thought it reasonable to believe that this group would thrive on advance organizer lesson methodologies of expository instruction.

Type 3 was much higher on spatial restructuring, rule discovery, and logical reasoning tests. We thought lessons for this group should emphasize discovery and 3-dimensional thinking.

Type 2 was strongest on tests using spatial orientation but not spatial restructuring. These children should enjoy engineering projects where general directions and guidelines are given.

A thought process of particular interest required the ignoring of standard word meanings in logical reasoning in the Nonsense Syllogisms test. Type 1 students appeared to use the words as meanings while Type 3 were able to use the words as objects.

Tests on which Type 1 scored substantially above Type 3 were Hidden Patterns and Identical Pictures. These tests requiring rapid recognition of pre-organized information.

Tests on which Type 3 scores substantially above Type 1 were Nonsense Syllogisms, Form Board, Paper Folding and Surface Development. These tests required spatial restructuring and reasoning using words as if they were objects.

We therefore concluded that "spatial ability" is not a single entity. Students either have three-dimensional transformational excellence or outstanding two-dimensional recognition ability – but not both. Our gifted population seemed to have individuals that have one or the other preferred mode of thinking. Where do the "other kind of children" fall? In my experience, they have been entirely Type 3.

We need to think deeply about the way we teach. Does the word "Inquiry" mean the same to all our teachers? To all our students? Are some forms of inquiry implemented with advance organizers while others are focused on discovery of pattern relationships. If we are to be a safe haven for this "other kind of child" then at least some of our instruction should provide opportunities in three-dimensional spatial restructuring and invention.

Appendix #3: Suitable Lessons

1. We want children to write.
    Let them write about physics.

2. We want them to learn mathematics.
    Let them discover it.
    Pose clever challenges that incorporate target concepts.

3. We want them to become rational citizens.
    Let them engage in scientific discourse.

Language Arts Selections To Start With

A literature selection suited to this type of developing interest must contain correct and believable physics, and yet it must facilitate "what if's " about the physical phenomena. Sociological and psychological aspects of the setting must be only peripheral elements in the story. "Feeling tone", so frequently uppermost in a literature teacher's mind, must not be the only focus of analysis questions.

The type of science fiction story advocated here is one which maintains conventional physics in all but one parameter. Twisting one law of physics with respect to all of the conventional laws allows the student to contemplate "what if .. ?" in terms of logical consequences. Please notice I am not advocating the type of literature called science fiction which treats the laws of physics as some magical or metaphysical construct where all or most of the laws of conventional physics are either violated, ignored, or "time-warped."

An excellent collection of stories edited by Isaac Asimov is entitled *Where Do We Go From Here?* Not all of the selections in this volume meet our criteria, but many do. The selection which has been most consistently successful is *The Big Bounce* by Walter Tevis. It is about a ball so elastic that once it started bouncing (how does this actually happen?), each bounce would be higher than the last. The special joy of the story is the developing awareness that the process violates the second law of thermodynamics – revealed in the surprise ending where, finally, the ball's heat was all used up as energy for bouncing, so it froze and shattered before it could do any damage.

In the process of writing the book, Asimov wrote a short commentary at the end of each story concerning the science concept violated and the scientific principles against which the action was set. He also posed questions at the end of each commentary which are worthy of a literature teacher's consideration. They require a student to write as clear and well composed an answer as any "humanistic" question the teacher might assign. They require analysis, synthesis and evaluation which go well beyond the information presented in the story – in keeping with the goals and objectives of curriculum tasks for the gifted. Most importantly, the questions do not violate the interests of this particular type of student. They facilitate thought about topics of interest to this "other kind of child."

The questions posed by Asimov at the end of *The Big Bounce* are:

1. *What is the stand of the U.S. Patent Office on "perpetual motion" machines? All of these, incidentally, break either the first or second laws of thermodynamics. Why?*

2. *The ocean has vast quantities of heat in it. Even a polar ocean does. Why do ocean vessels have to burn fuel? Why can't they just use the heat of the ocean water over which they pass?*

3. *In view of the first law of thermodynamics, where does the tremendous energy radiated by the Sun and all other stars come from?*

4. *Look up the history of some perpetual motion machines of the past. What was the "catch" in each case? Why wouldn't they work? Were some of these outright hoaxes?*

5. *The laws of thermodynamics are based on the general experience of scientists. They have never observed the laws to be broken. Scientists, however, merely observe their own section of the universe and their own general type of environment. What about outer space ten billion light-*

years away? What about the center of the Sun? How sure can we be that scientific laws are the case everywhere under all conditions?

6. *Suppose a scientist discovered some easily produced phenomenon which seemed to defy either the first or second law of thermodynamics. Should he assume at once there was some mistake and forget the whole thing? Should he instantly proclaim the phenomenon and declare the laws broken? What would you do?*

 Does a literature teacher have to have an extensive knowledge of science and mathematics in order to grade these questions? I think not. These questions are probably general enough in nature to be graded by most literature teachers.

Another story that is very popular with this "other kind of child" comes from the same collection by Asimov. The story is entitled *The Holes Around Mars* and is by Jerome Bixby. After detecting strange gouges in rocks, leaves, and hills which all seem to line up, the spaceship crew discovers (in an amusing manner) that the planet has a moon traveling just four feet above the ground. Of course such a moon would have to travel at impressive speed to keep at that altitude. Since the other moons of Mars are named Deimos and Phobos, the newly discovered moon is named "Bottomos."

After pointing out the scientific flaws in the story in his short commentary, Asimov posed these questions:

1. *What is the orbital velocity of the two known Martian satellites? If you divide the escape velocity from the Martian surface by the square root of two you would obtain the orbital velocity for a satellite revolving in a circular*

orbit in the neighborhood of the Martian surface? How much is that in miles per hour. If the satellite were of ordinary granite, what would its kinetic energy be in comparison with that of a .45 bullet shot out of a revolver?

2. *If the third satellite were neutronius, how much would it weigh? What would be its kinetic energy at the orbital velocity?*

3. *Our explorers in The Holes Around Mars find that the holes are in a straight line; that is, if you sight through a hole in one place you would see another hole in another place far away. But could you? Is the third satellite traveling in a true straight line or is it following the curvature of the Martian surface? How much does the Martian surface curve; that is, in the space of one mile, how far does the Martian surface drop? Calculate how far apart two holes must be so that when sighting through one you can no longer see the other.*

4. *According to the story, it would seem that the third satellite follows precisely the same path each time it circles the planet. Yet if the orbit were at an angle to the Martian equator, it would follow a different path each time around, for Mars will have rotated part-way during the time of one orbit. What could the orbit look like on a flat map of the Martian surface?*

5. *If the third satellite were following a path exactly along the Martian equator, the rotation of the planet wouldn't matter and the little satellite would follow its own track each rotation. (Why?) Is it likely, though, that its orbit would be lined up exactly with the Martain equator? What about the orbits of Phobos and Deimos in this respect? Suppose the third satellite's orbit varied from the line of the equator so slightly that it moved four inches north of the equator at*

one extreme of its orbit and four inches south at another. What would that do to the holes?

Although some of the stories in Asimov's book *Where Do We Go From Here* were not received well by "this other kind of child", most of them were. Another story in the collection, *A Subway Named Mobius* by A. J. Deutch, is a delightful story which can be followed by having the students actually construct Mobius strips (one-sided pieces of paper) out of paper and tape. This causes one to contemplate the way two-dimensions can warp through three-dimensions.

And so we have the challenge: can instructional methods be developed which will maintain the interest of "this other kind of child"? Clearly, the existing language arts instructional methods are adequate for many students. We have produced a generally literate populace. Our concern is that a small but rather important group of students seems to have been overlooked in our development of instructional alternatives. It is hoped that by understanding the cognitive style and developmental time-frame of this group of students, educators will be able to develop appropriate material for them.

A Lens To Craft Mathematics And Science Lessons

We have some ideas about how to tailor Language Arts lessons to the needs of "this other kind of child." What about Mathematics? What is it about the way our country's teachers teach drill and practice workbook math that so alienates this other kind of child? How can we re-craft our lessons so they learn the math ideas they need and yet are energized by doing so?

Primary Age Students

Our research suggests an organizing lens. Some children are quite good at thinking about 3-dimensional objects and their rotations. Where in school do we explore 3-dimensions?
We have an engaging third grade classroom that allows children to build and design with materials such as Kapla blocks, croutons, and paper origami. Photos of the structures the students have created are impressive. Projects seemed to inspire creative construction wherein students explored geometry and physics. Balancing and fitting polygons. Ratio and proportion. Blocks were created in ratios of: 1,3,15 leading to new ways of understanding structures. As I looked at the work, several students came into the room describing with animation their love of the spatial projects available to them. What a wonderful setting. It seems to be serving the needs of our range of spatially able children.

Intermediate Age Math and Science.

Moving up to Intermediate age students, we need to imbed more rigorous, labeled mathematical ideas. Exploration is still essential but with it must come some recognizable adult concepts. Students in intermediate years are not yet very good with calculations but they can learn to graph. They can extract meaning from graphing their own and the class data and choosing their own confidence intervals.

So, when do these students engage? What causes them to actually listen to each other? If we can contrive a setting where the teacher steps back from being the purveyor of wisdom and there are no textbooks, answer book or internet, then students may turn to the findings of other students to get feedback on their own correctness. It is in their own self-interest to listen to other students. They want to know if they are on the right track. Real conversations. Deep engagements. Reasoning and making sense of uncertainties. Meaningful dialogue – about mathematics. Is that possible? I think so. I watch it happen on a daily basis. It is like playing music. You have to know how much information to present. When to step back. When to let students generate their own data. When to have them compare and contrast their data. When to add a tidbit of information and when to leave levels of uncertainty dangling a bit. A teacher has to listen, really listen to these students, to determine how the "music" is flowing. Rich mathematical discourse is possible for "this other kind of child."

Rich dialogue. That's the classroom tone I set whether I teach mathematics or science. We don't use textbooks and we don't have answer books. We have each other. Our judgment of our own work comes from a comparison and reasoned discourse with others. I, as the teacher, have to decide on the concept I want students to acquire. Then I have to devise a project to draw them along to acquire, even discover that idea.

Right now I want to teach them to graph. Not a bar graph, but a linear equation graph. Fourth graders. I want 10 year olds to make meaning from a graph before they ever encounter it in pre-algebra math class. Meaning first – algebraic manipulation second.

So, how do I do that? I introduce triple-beam balances. Fourth graders feel very grown up using such fancy equipment. (Coat hanger balances worked pretty well in years past.)
And 10 ml. graduated cylinders - they are very grown up pieces of equipment too. So, how much does a 10ml. grad weigh? Every pair of students goes to a work table and weighs their graduate. I have 16 students in a class so that makes 8 pairs of students. Eight data points. The class gathers and each and every student takes out their notebook (journal) and writes, "class data." Then we all write, Lab 1,2,3,4,…8 on a function-table chart. Lab #1 reports the weight of their grad: 35.6. Lab #2 reports: 35.8. Lab #3; 34.9. And so on around the room. Soon we have eight data points. I ask, "What is our conclusion – What do graduates weigh?" "Is there a range?" "A trend?" "Were all the grads the same brand?" "Did any have a collar?" What are possible uncontrolled variables?

So now, what will water weigh? After explaining a meniscus and the ml. lines inscribed on the glass of the graduated cylinders, students are ready to weigh successive milliliters of water. Although they will work in pairs, each person will prepare a function-table with space to record the weight of each successive ml. of water. Each person will record, for themselves, the weight of each ml. of water as they go along. The pairs of students go to work taking turns filling water to the next ml. line or running the riders to determine the weight. Everyone works. Everyone handles a balance. Everyone adds liquid. Every single student gets a "feel" for what we are doing. No standing and watching others. By the end of the period most students have data for 8 to10 ml.

Students rush into class the next day and open their notebooks.
We write, "class data."
Each student lists lab numbers down the left hand side of their
page and draws twelve blank columns filling the rest of the page.
The lab pairs report their findings of the weight of 1ml, 2ml, etc.
Students alternate so by the time all eight labs have reported, every
student in the room has contributed. But what about the twelfth
column? As each lab shares their data points we together subtract
(calling out loud so we hear each other's calculation and come to
agreement) to find how much the value has gone up. Lines wobble
a bit with the numbers increasing by .8, .9, 1 or even 1.1 each time.
At the end of each lab's report, we ask, " What is the trend?"
"What would you say?" "Don't average the numbers – just look
for the main pattern." The twelfth column is for the trend. As
subsequent labs report, we notice whether they agree with previous
labs. After all labs finish we say: "Conclusion." Well, the "trend"
column shows it is usually pretty close to one. The world might be
telling us that the weight of water is one. One gram per ml! It
looks like water pretty much weighs one gram per ml!   No
textbooks. No memorized answers. No boring show-and-tell
reporting. Everyone reports. Everyone listens. Everyone is
engaged and interested. They want to know if they are getting
close to "right." Everyone seems thrilled when we agree. Not
perfectly, but then this is not a perfect world. We have
uncontrolled variables, Sticky balances. Drops of water.
Imperfect meniscuses. But there seems to be an underlying pattern
we can all agree on.

Well yes, some of their parents told them they were "correct."
Water weighs one gram. But the precision isn't the point. The
mathematical discourse is. And now we graph our personal data.
Weight on the y-axis, ml. on the x-axis. We plot the weight of
water and see that the line goes over one ml. and up one gram. Up
one, over one. Slope is rise over run. Our water line has a slope of
about 45°, No mention of linear equations. No mention of y-

intercept. Just the introduction of the idea of slope generated from their own data.

But now, what about "Green Poison?" I have never told anyone over the years that it is salt water with green food color. I call it poison – (enough salt water can hurt you, you know). The next day students scurry to lab with medicine cups of "Green Poison." Each student again prepares a function-table of ml. of liquid to receive the weight reading of each ml. of the new liquid. The following day students come to class and excitedly prepare their journals to record class data about "Green Poison". "Is it really poison?" "Will it be the same as water?"

Lab #1 reports. After each ml. the class yells out the difference in weight from the previous reading. Five ml. equals 40.2. Six ml. equals 41.4. Oh my goodness! It is going up 1.2. Or 1.1. Or 1.3. Anyway it is heavier than the water data. Once every lab has reported and the "Oh's and Ah's" are over, each student adds their personal data to their individual graph. Well, look at that! Green Poison is heavier (pulled down by gravity) but the graphed line goes higher. Yes that makes sense. More mathematical discourse. More meaning making. And every single student has the opportunity to contribute. We need each other to compare our data and find the pattern.

This Other Kind Of Child

1/3/17

Lecture

We took water data on december 13th. We found that water t=1
Green poison
Class Data

Lab	Brand	0	1	2	3	4	5	6	7	8	9	10	
1													
2	Tekk	28.5	29.5	30.6	31.8	33	34.2	35.4	36.5	37.7	38.8	39.4	1.2-1.1
3	Kiroa	40.1	41.7	43	44.3	46.1	46.8	47.5	48.8	49.9	51.1	52.4	1.3
4	Sibota	24.5	25.5	26.6	27.8	29.1	30.1	31.5					1.2
5	Sibota	24.6	25.5	26.5	27.8	29.1	30.1	31.3	32.5	33.7	31.9	36	1.2
6	Sibota	35	35.8	37.2	38.3	39.5	40.8	41.8	43.1	44.1	45.4	46.6	1.3
7	TC	33.5	34.5	36.5	37.5	38.1	39.2	40	41.8	42.8	43.4	45.1	7
8	Sibota	36.3	37.1	38	39.1	40.3	41.5	42.1	44	45.4	46.8	47.8	1.2-1.4

Conclusion:
Green poison ranges from 1.1→1.4
mostly 1.2-1.3

Now what about "Blue Poison?" I set out 99% rubbing alcohol stained blue with food color. Of course, I don't tell them what it is. Mystery makes it more fun. Is it the same as green? How will you know? Some students try to smell it but I refocus them on whether it will weigh the same or different. They are all eager to engage. After a day of work, they scurry in to report their data. Milliliter #1 weighs 46.4. Milliliter #2 weighs 47.2. Milliliter #3 equals 48 even. I subtract aloud and students shout out. It is going up less! About .8. How exciting! The tone of the class is excitement! Over subtraction problems! Isn't that remarkable! Fourth grade students are enthusiastic over subtracting and getting .7 or .8. Mathematics can be authentic. Mathematics can be engaging. Reporting numerical findings can go beyond show-and-tell to become deep engagement in discovering mathematical regularities.

Students eagerly enter data points for the "Blue Poison" on their graphs. Aha! "I knew the line would be lower." "The slope would not be as steep."

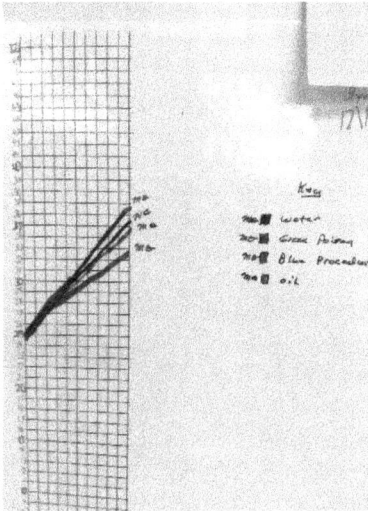

We repeat the process with cooking oil. For the faster students I make available "Yellow Poison" (70% rubbing alcohol) for extra credit.

So now the graphs are laminated and proudly taped to the wall of each student's work station. One last task. Using the information on your graph, make a marble float in the middle of a medicine vial. Well now, looking at the graph, the slope of the green is steeper so that means it is heavier and might sink. The slope of the blue is shallower so that is lighter and might float. The oil line is right next to the water line, and oil is thick and sticky so that might be the marble. "What is your recipe?" Class data. "How did you do that?" Meaningful discourse. Rich conversations. And an understanding of slope, its graphical representations and meaningful use of those representations. This process engages this "other kind of child."

Middle School and High School mathematics.

Our students have painted an Escher tessellation on the stairway walls of our Middle School.

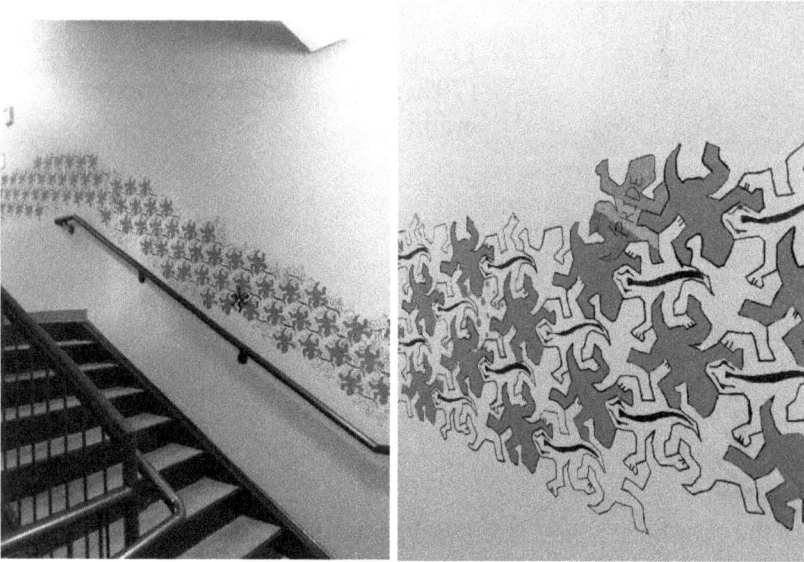

Spatial activities are included in the math program but increasing emphasis is placed on preparing students to flow into future math courses found in schools they will next be attending.

The next challenge for me is to explain ways to teach this older child. The Junior High School and High School age. As the concepts of Algebra and Geometry become more specific and formalized, must we regress to "show and tell" instructional methods? Not necessarily. I have worked on this problem for several decades and have come up with projects our alumnae fondly remember.

They will be discussed in my next *Raising Interesting People* publication.

Dr. Meredith Olson, known affectionately as Doc "O" to her students, has taught elementary, middle school and high school math and science in Seattle for nearly 60 years. Her primary goal is in improvement of pre-college engineering education. By going to lab to work on contraptions every day, her students come to understand properties of the mechanical world.

"It has been a long and interesting trip. Studying some metallurgy in grad school. Evening classes. After a full day of high school teaching. Consulting for JPL as the Mars Pathfinder Educator. Weekends. Working in the summer with UNESCO in Zimbabwe, Kenya, and Uganda. Teaching dozens of weekend and week-long summer teacher workshops in South Carolina and Montana. Being a consultant and curriculum designer for Health and Physiology education in Washington, Oregon, Idaho, Montana, and Alaska. Being a summer adjunct University instructor for more than 20 years in Seattle, Idaho and Montana. Teaching teachers. Teaching students every day, every year for 59 years. Observing how learning happens. Becoming aware when real learning isn't happening. When it is just "show". When it is just teacher–pleasing to get a grade. To get a credit. To get a university degree."

See Dr. Olson's open letter outlining her philosophy of lesson design, available on the JPL website - Exploring Preface pp 11-13
http://mars.jpl.nasa.gov/education/modules/GS/GS07-19_preface.pdf

Dr. Olson believes that children must construct their own understanding from active design and assemblage of contraptions. By testing, failing, remodeling, and trying again, we come to see the structure when we look. By carefully examining materials we have, we may perceive how to use them in new and unexpected ways. Children begin to understand the engineering process. Besides, it is fun!

www.ingramcontent.com/pod-product-compliance
Lightning Source LLC
Chambersburg PA
CBHW062027040426
42447CB00010B/2174